3 Connecticut SBAC Grade 4 Math Practice Tests

Full-Length Test Prep with Detailed Answer Explanations

Dr. A. Nazari

Grade 4 Math: 3 Practice Tests

Hey there, quick checker! 

Three tests. That's all it takes to **see where you stand** and **build your confidence** fast.

✅ Short and focused — perfect for a quick review.

✅ Find your strengths in just three rounds.

✅ Fast results, real progress!

Ready for a quick check? Let's go! 

> **Three quick tests and you'll know exactly what to study next!**

📋 How to Use This Book 📋

Three tests, one clear plan — here's how to make them count.

📖 What's Inside

- **3 Practice Tests** — quick, focused tests covering all Grade 4 topics
- **Answer Key** — detailed answers and explanations at the back
- **Reference Pages** — math symbols and multiplication table you can use during tests
- **My Test Tracker** — record your scores and see your growth at a glance

📋 How to Take a Practice Test

1. **Set up:** Find a quiet spot. Gather pencils, eraser, and scratch paper.
2. **Time it:** Set a timer (or ask a grown-up to). Try to finish without rushing.
3. **Work through it:** Answer every question. Skip hard ones and come back later.
4. **Check your work:** Use any extra time to double-check your answers.
5. **Score it:** Use the Answer Key. Write your score in the Test Tracker.

✅ Multiple Choice

Pick the best answer from the choices given. Read ALL options before choosing — the first one that looks right isn't always the best!

✏️ Short Answer

Solve the problem and write your answer. Show your work — even if you get the answer wrong, partial credit counts in many real tests!

> ⓘ **Quick Tip:** With only 3 tests, make each one count. Review every missed question before moving to the next test!

# Test-Taking Tips

Quick strategies to help you ace all 3 tests!

Before You Start

- Get a good night's sleep and eat a healthy snack.
- Have your supplies ready: pencils, eraser, scratch paper.
- Take three slow, deep breaths — you've got this!
- Remember: just 3 tests, so give each one your full focus.

During the Test

1. **Read each question twice.** Underline key words like "how many more," "product," or "estimate."
2. **Show your work.** Write out each step — it helps you catch mistakes.
3. **Use scratch paper.** Line up digits carefully for multi-digit problems.
4. **Skip and return.** Stuck on a question? Star it and move on. Come back with fresh eyes.
5. **Check with estimation.** Does your answer make sense? A quick estimate can catch big errors.
6. **Use your time wisely.** Don't spend too long on one problem.

Multiple Choice Tricks

- Read **all** choices first.
- Cross out answers you know are wrong.
- Plug your answer back into the problem.
- When in doubt, eliminate and guess — never leave it blank!

Watch Out For...

- Rushing through without reading carefully.
- Mixing up × and + in word problems.
- Forgetting to regroup when adding or subtracting.
- Not simplifying fractions when asked.
- Skipping the "check your work" step.

 **"** Three tests is all you need for a quick check-up! Focus on each one, learn from your mistakes, and you'll be amazed how much you grow. **"**

# What You'll Need

Grab these supplies before you start each test.

Pencils

#2 pencils are perfect!

Eraser

Clean erasing = cleaner work

Scratch Paper

For all your work!

Quiet Space

Like a real test room

Timer (Optional)

Practice pacing yourself

Confidence!

Three tests, three wins!

✅ You CAN Use

- Pencils and eraser
- Scratch paper
- The reference pages in this book
- A ruler (for measurement questions)

❌ You CANNOT Use

- Calculator
- Phone or tablet
- Other books or notes
- Help from anyone else

👥 For Parents & Teachers

This compact 3-test set is ideal for a quick skills check. Simulate real test conditions when possible. After each test, review missed questions together and focus on **growth** ("You improved on fractions this time!") rather than the raw score.

X¹ Math Reference Sheet X¹

You may use this page during your practice tests!

Symbol	Name	Meaning
$+$	Plus	Combine amounts
$-$	Minus	Find the difference
$\times$	Times	Multiply (repeated groups)
$\div$	Divide	Split into equal parts
$=$	Equals	Same value on both sides
$>$ $<$	Greater / Less Than	Compares two values
$\frac{a}{b}$	Fraction	a parts out of b equal parts
.	Decimal Point	Separates wholes from parts
$\angle$	Angle	Measured in degrees

Key Math Words

- **Factor** — a number you multiply
- **Product** — answer from multiplying
- **Quotient** — answer from dividing
- **Remainder** — left over after dividing
- **Numerator** — top of a fraction

- **Denominator** — bottom of a fraction
- **Equivalent** — equal in value
- **Perimeter** — distance around
- **Area** — space inside
- **Estimate** — a close, rounded guess

🔍 Word Problem Clue Words

Add (+)   in all, altogether, total, combined, sum, increase

Subtract (−)   how many more, how many fewer, difference, left, remain

Multiply (×)   each, every, per, times as many, groups of, product

Divide (÷)   split equally, shared among, divided into, per group

Find more at
ViewMath.com/CT-Grade4

Multiplication Table

×	1	2	3	4	5	6	7	8	9	10	11	12
1	1	2	3	4	5	6	7	8	9	10	11	12
2	2	4	6	8	10	12	14	16	18	20	22	24
3	3	6	9	12	15	18	21	24	27	30	33	36
4	4	8	12	16	20	24	28	32	36	40	44	48
5	5	10	15	20	25	30	35	40	45	50	55	60
6	6	12	18	24	30	36	42	48	54	60	66	72
7	7	14	21	28	35	42	49	56	63	70	77	84
8	8	16	24	32	40	48	56	64	72	80	88	96
9	9	18	27	36	45	54	63	72	81	90	99	108
10	10	20	30	40	50	60	70	80	90	100	110	120
11	11	22	33	44	55	66	77	88	99	110	121	132
12	12	24	36	48	60	72	84	96	108	120	132	144

Quick Reminders

Multiply: Find one number on the left, the other on top. Where they meet = your answer.

Divide: For 96 ÷ 8, find 96 in the 8-row. The column header tells you the answer: 12!

Factor pairs: Any number in the table can be written as row header × column header.

📈 My Test Tracker 📈

Track your 3 tests and watch yourself improve!

Name: _______________________________________

Test #	Date	Score	How I Feel
1			
2			
3			

❝ Three tests, three chances to shine! Compare each score to your last one — even one extra point means you learned something new! ❞

Quick progress is still progress! Three tests gave you a clear picture. Now you know exactly where to focus next. Be proud of every step forward!

 Table of Contents

Here's what we'll explore together!

 Let's learn and have fun!

1

Practice Test 1

 30 Questions

---

✏️ Before You Start ✏️

- ✓ **Read each question carefully** before choosing your answer.
- ✓ **Show your work** on scratch paper when you need to.
- ✓ **Skip hard questions** and come back to them later.
- ✓ **Check your answers** when you're done.
- ✓ **Take your time** — there's no rush!

 ⭐ You've Got This! ⭐ 

Do your best and show what you know!

1. A box has 5 pencils. Another box has 7 times as many pencils. How many pencils are in the bigger box?

Your Answer:

2. There are 45 students split equally into 5 teams. Each team needs 3 jump ropes. How many jump ropes are needed in total?

Your Answer:

3. I am thinking of a number between 40 and 50 that has 7 as a factor. What is the number?

Your Answer:

4. Is 41 prime or composite? Explain how you know.

Your Answer:

5. In 307,416, which digit is in the hundred-thousands place?

(A) 7

(B) 4

(C) 3

(D) 0

6. Which comparison is **INCORRECT?**

(A) $745,000 > 74,500$

(B) $600,001 > 599,999$

(C) $410,300 < 410,030$

(D) $325,400 = 325,400$

Find more at
ViewMath.com/CT-Grade4

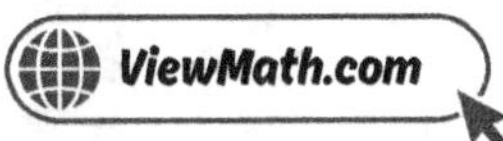

7. If $3 \times 10 = 30$, what is 3×100?

(A) 30

(B) 300

(C) 3,000

(D) 3,030

8. Which is the best estimate for $387 + 425$ using rounding to the nearest hundred?

(A) 700

(B) 800

(C) 900

(D) 750

9. List the partial products for 7×428 and find the total.

Your Answer:

10. Which of the following best describes the standard algorithm for 78×43?

(A) Multiply 78×3, then multiply 78×4, and add those two rows

(B) Multiply 78×3, then multiply 78×4 (shifted one place left), and add

(C) Multiply 78×43 all at once in one step

(D) Add $78 + 43$ and then double the result

11. Which fraction is the greatest?

(A) $\frac{1}{6}$

(B) $\frac{1}{3}$

(C) $\frac{1}{2}$

(D) $\frac{1}{4}$

12. A pan of brownies is cut into 8 equal pieces. Leo ate $\frac{1}{8}$, Sam ate $\frac{2}{8}$, and Kai ate $\frac{1}{8}$. What fraction of the brownies did they eat altogether?

(A) $\frac{3}{8}$

(B) $\frac{4}{8}$

(C) $\frac{5}{8}$

(D) $\frac{6}{8}$

Find more at
ViewMath.com/CT-Grade4

13. What is $\frac{5}{12} - \frac{2}{12}$?

(A) $\frac{3}{0}$

(B) $\frac{7}{12}$

(C) $\frac{3}{12}$

(D) $\frac{2}{12}$

14. What is $6\frac{3}{5} - 4\frac{4}{5}$? (Regroup as needed.)

(A) $2\frac{1}{5}$

(B) $1\frac{4}{5}$

(C) $2\frac{4}{5}$

(D) $1\frac{3}{5}$

15. What is $6 \times \frac{1}{4}$, written as a mixed number?

(A) $1\frac{1}{4}$

(B) $1\frac{3}{4}$

(C) $1\frac{2}{4}$

(D) $2\frac{1}{4}$

16. A dime is $\frac{1}{10}$ of a dollar, and a penny is $\frac{1}{100}$ of a dollar. How much of a dollar is 3 dimes and 8 pennies combined?

(A) $\frac{30}{100}$

(B) $\frac{38}{100}$

(C) $\frac{11}{100}$

(D) $\frac{308}{100}$

17. Write 0.25 as a fraction.

(A) $\frac{25}{10}$

(B) $\frac{25}{100}$

(C) $\frac{2}{5}$

(D) $\frac{2}{10}$

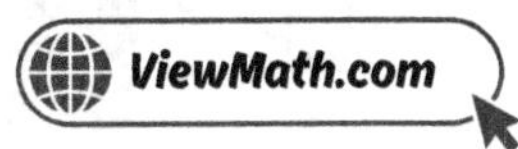

18. Two swimmers raced. Swimmer A finished in 32.4 seconds. Swimmer B finished in 32.40 seconds. Who won?

(A) Swimmer A, because $32.4 < 32.40$.

(B) Swimmer B, because $32.40 > 32.4$.

(C) It was a tie, because $32.4 = 32.40$.

(D) Not enough information.

19. How many meters are in 1 kilometer?

(A) 100 *m*

(B) 1,000 *m*

(C) 10 *m*

(D) 10,000 *m*

20. Look at the diagram below. A track is 400 m around. The runner has completed 3 laps. How many kilometers has the runner covered?

400 *m per lap*

Laps completed: 3

Your Answer:

21. A square room has sides of 10 feet. What is the area of the room?

(A) 40 *sq ft*

(B) 20 *sq ft*

(C) 100 *sq ft*

(D) 1,000 *sq ft*

22. What is the perimeter of a rectangle that is 9 cm long and 4 cm wide?

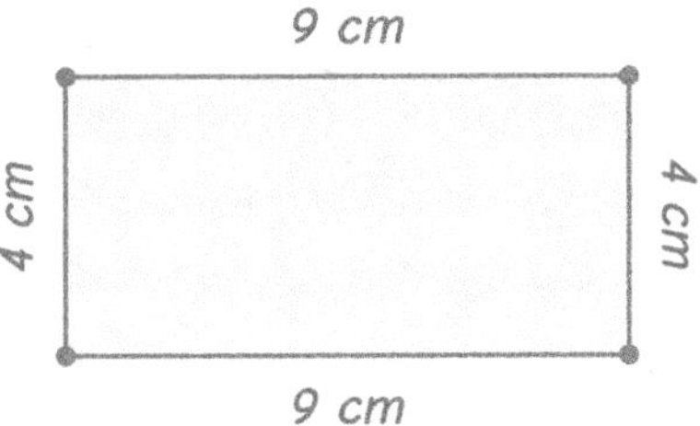

(A) 36 cm

(B) 26 cm

(C) 13 cm

(D) 52 cm

23. A number line for a line plot measures in $\frac{1}{2}$-cup intervals. Which values should appear on the number line between 0 and 1?

(A) $\frac{1}{4}$

(B) $\frac{3}{4}$

(C) $\frac{1}{2}$

(D) $\frac{2}{3}$

24. Which angle measure is obtuse?

(A) 45°

(B) 89°

(C) 110°

(D) 180°

25. Angle A measures 45° and Angle B measures 45°. Together, what type of angle do they form?

(A) Acute

(B) Right

(C) Obtuse

(D) Straight

Find more at
ViewMath.com/CT-Grade4

26. Points C, D, and E are all on the same line. How is segment $\overline{CD}$ related to segment $\overline{DE}$?

(A) They are the same segment

(B) They are parts of the same line

(C) They are parallel

(D) They are perpendicular

27. Which angle type has the LARGEST measure of the four types?

(A) Acute

(B) Right

(C) Obtuse

(D) Straight

28. True or false: Parallel lines can intersect if they are made long enough.

Your Answer:

29. A square is BEST described as:

(A) A rectangle with 4 equal sides

(B) A rhombus with no right angles

(C) A trapezoid with 4 equal sides

(D) A triangle with 4 right angles

30. How many lines of symmetry does an equilateral triangle (all sides equal) have?

(A) 1

(B) 2

(C) 3

(D) 0

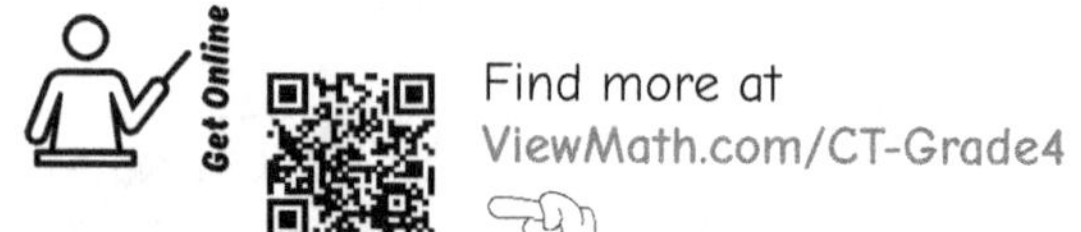

 End of Practice Test 1 

Great job finishing the test!

My Score

I got _____________ out of 30 questions right.

Check your answers in the **Answer Key** at the back of the book.

Review any questions you missed. That's how we learn!

Check Your Score Online!

Visit **ViewMath Academy** to enter your answers and see which topics you need to review. You can also explore lessons, take quizzes, track your scores, and save your progress!

viewmath.com/score/4.1.CT.01

Or go to viewmath.com/score and enter code: 4.1.CT.01

2

Practice Test 2

 30 Questions

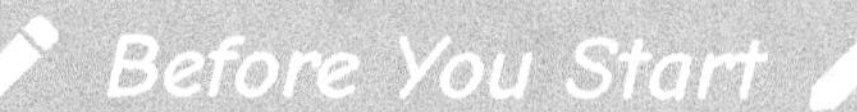 Before You Start

- ✓ **Read each question carefully** before choosing your answer.
- ✓ **Show your work** on scratch paper when you need to.
- ✓ **Skip hard questions** and come back to them later.
- ✓ **Check your answers** when you're done.
- ✓ **Take your time** — there's no rush!

 You've Got This! 

Do your best and show what you know!

1. How many total dots are in the array below? Write a multiplication equation for the array.

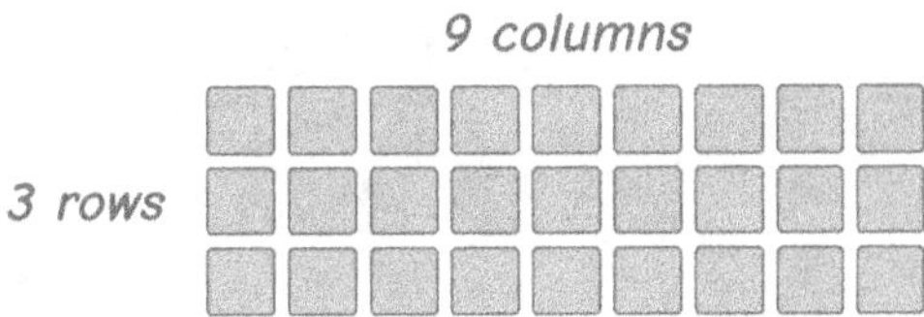

Your Answer:

2. A school ordered 7 boxes of crayons. Each box has 24 crayons. The teacher gave away 15 crayons. How many crayons does the school have now?

(A) 168

(B) 153

(C) 163

(D) 152

3. Name a number between 1 and 20 that has exactly 4 factors.

Your Answer:

4. Which number is a multiple of both 3 and 5?

(A) 20

(B) 25

(C) 30

(D) 35

5. What is the value of the digit 5 in 850,293?

(A) 5

(B) 500

(C) 5,000

(D) 50,000

6. *The population of Town A is 183,400 and Town B is 183,040. Which town has more people?*

   (A) Town B   (B) They are equal

   (C) Town A   (D) Cannot tell

7. *Look at the diagram below. The same digit 9 appears in two positions. How many times bigger is the value at Position A than at Position B?*

TTh	Th	H	T	O
9	2	**9**	4	1

     ↓   ↓
     A   B

Your Answer

8. *Round 347 to the nearest ten.*

   (A) 300   (B) 340

   (C) 350   (D) 400

9. *Each page of a photo album holds 7 photos. The album has 215 pages. How many photos can the album hold?*

Your Answer

10. *A box holds 36 crayons. A school orders 52 boxes. How many crayons did the school order?*

   (A) 1,782   (B) 1,872

   (C) 1,882   (D) 1,972

11. *True or false:* $\frac{2}{3} > \frac{5}{6}$.

Your Answer:

12. *Which is a correct way to decompose $\frac{7}{8}$?*

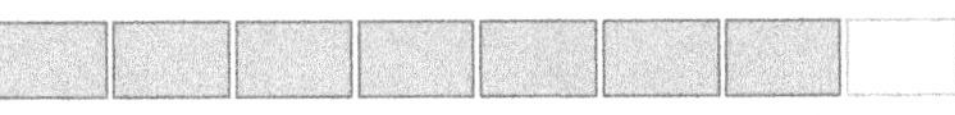

(A) $\frac{4}{8} + \frac{4}{8}$ (B) $\frac{3}{8} + \frac{4}{8}$

(C) $\frac{5}{8} + \frac{3}{8}$ (D) $\frac{2}{8} + \frac{6}{8}$

13. *Which addition has the same result as $\frac{5}{8} - \frac{2}{8}$?*

(A) $\frac{1}{8} + \frac{1}{8}$ (B) $\frac{1}{8} + \frac{2}{8}$

(C) $\frac{2}{8} + \frac{1}{8} + \frac{1}{8}$ (D) $\frac{2}{8} + \frac{2}{8}$

14. *What is $3\frac{7}{8} - 1\frac{5}{8}$?*

(A) $2\frac{3}{8}$ (B) $1\frac{7}{8}$

(C) $2\frac{2}{8}$ (D) $1\frac{2}{8}$

15. What is $5 \times \frac{1}{3}$, written as a mixed number?

  (A) $\frac{5}{3}$ (B) $1\frac{2}{3}$

  (C) $2\frac{1}{3}$ (D) $1\frac{1}{3}$

16. What fraction must be added to $\frac{3}{10}$ to get $\frac{45}{100}$?

  (A) $\frac{15}{100}$ (B) $\frac{42}{100}$

  (C) $\frac{5}{100}$ (D) $\frac{75}{100}$

17. Look at the 10×10 grid below. What decimal represents the shaded part?

  (A) 0.035 (B) 0.35

  (C) 3.5 (D) 35.0

18. Emma ran the 100-meter dash in 14.5 seconds. Mia ran it in 14.38 seconds. A student says Emma was faster because $14.5 < 14.38$. Is the student right?

  (A) Yes, because $5 < 38$.   (B) No. $14.5 = 14.50 > 14.38$, so Mia was faster.

  (C) No. $14.5 < 14.38$, so Emma was faster.   (D) Yes, because 14.5 and 14.38 have different numbers of decimal places.

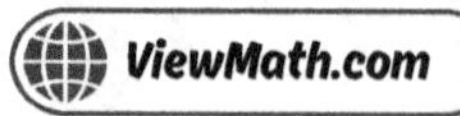

19. *The table below shows metric conversions. Use it to find how many millimeters are in 4 cm.*

Centimeters	Millimeters
1 cm	10 mm
2 cm	20 mm
3 cm	30 mm
4 cm	?

(A) 4 mm

(B) 14 mm

(C) 40 mm

(D) 400 mm

20. *Look at the bar model below. Find the total weight of both packages in grams.*

Total weight = ?

Package A: 4 kg	Package B: 1,500 g

Your Answer

21. A classroom floor is 9 m × 8 m. A bookshelf covers a 2 m × 3 m area. What area of floor is NOT covered by the bookshelf?

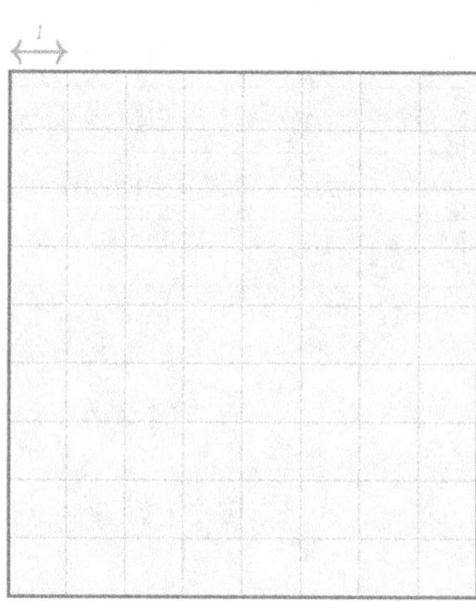

A) 66 sq m

B) 54 sq m

C) 48 sq m

D) 60 sq m

22. Look at the shape below made from two connected rectangles. What is the total perimeter?

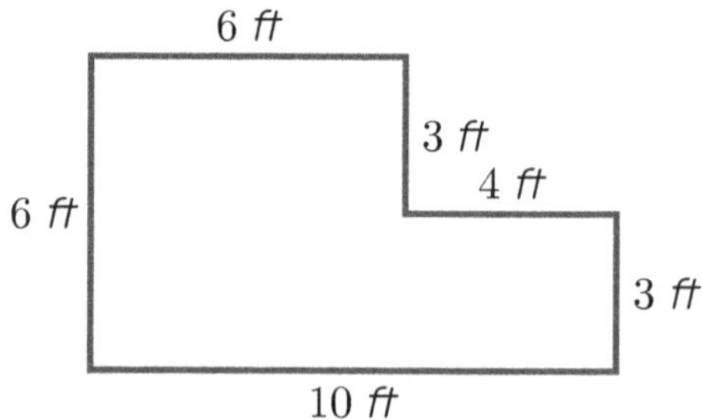

A) 28 ft

B) 30 ft

C) 32 ft

D) 36 ft

23. Lisa measures pencil lengths: 2 pencils at $\frac{5}{8}$ in, 3 pencils at $\frac{6}{8}$ in, 1 pencil at $\frac{7}{8}$ in. What is the longest pencil length measured?

A) $\frac{5}{8}$ in

B) $\frac{6}{8}$ in

C) $\frac{7}{8}$ in

D) $\frac{8}{8}$ in

24. A quarter turn (going one-fourth of the way around a circle) measures how many degrees?

A) 45°

B) 90°

C) 180°

D) 360°

25. Four angles meet at a point and together make a full 360° turn. Three of the angles measure 90°, 90°, and 80°. What is the fourth angle?

A) 80°

B) 90°

C) 100°

D) 110°

26. Maya draws a figure that has a beginning but no end. It goes on forever in one direction. What did she draw?

A) A line

B) A line segment

C) A point

D) A ray

27. Look at the angle below. Is it acute, right, or obtuse? Write your answer.

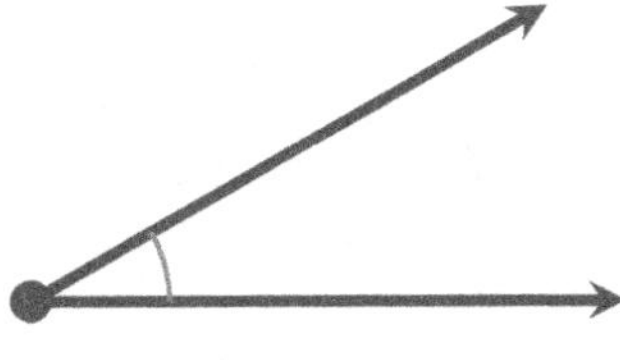

Your Answer:

28. Which of these real-world examples best shows perpendicular lines?

(A) Two lanes of a highway going straight ahead

(B) The top and side edges of a door

(C) Two train tracks running side by side

(D) The two sides of a ladder

29. How many sides does a quadrilateral have?

(A) 3

(B) 4

(C) 5

(D) 6

30. How many lines of symmetry does an equilateral triangle have?

Your Answer:

 # End of Practice Test 2 

Great job finishing the test!

☑ My Score

I got _____________ out of 30 questions right.

*Check your answers in the **Answer Key** at the back of the book.*

💡 *Review any questions you missed. That's how we learn!*

📊 Check Your Score Online!

Visit **ViewMath Academy** to enter your answers and see which topics you need to review. You can also explore lessons, take quizzes, track your scores, and save your progress!

viewmath.com/score/4.1.CT.02

Or go to viewmath.com/score and enter code: 4.1.CT.02

3

Practice Test 3

 30 Questions

✏️ Before You Start ✏️

- ✔ **Read each question carefully** before choosing your answer.
- ✔ **Show your work** on scratch paper when you need to.
- ✔ **Skip hard questions** and come back to them later.
- ✔ **Check your answers** when you're done.
- ✔ **Take your time** — there's no rush!

 You've Got This! 

Do your best and show what you know!

1. The number bond below shows a multiplication comparison. What number belongs in the top circle?

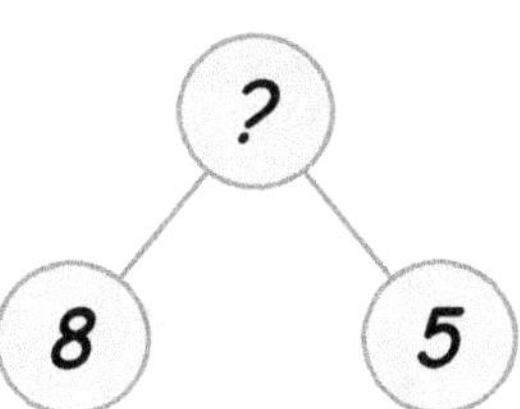

Hint: The top number is 8 times as many as 5.

Your Answer:

2. Sofia baked 48 cookies. She put them equally into 6 bags. She then bought 5 bags from a friend. How many bags does she have altogether?

(A) 11

(B) 8

(C) 13

(D) 43

3. Which number has exactly one factor pair?

(A) 12

(B) 9

(C) 7

(D) 16

4. I am a prime number between 45 and 55. What am I? (There may be more than one answer.)

Your Answer:

5. Look at the base-ten blocks below. What number do they show?

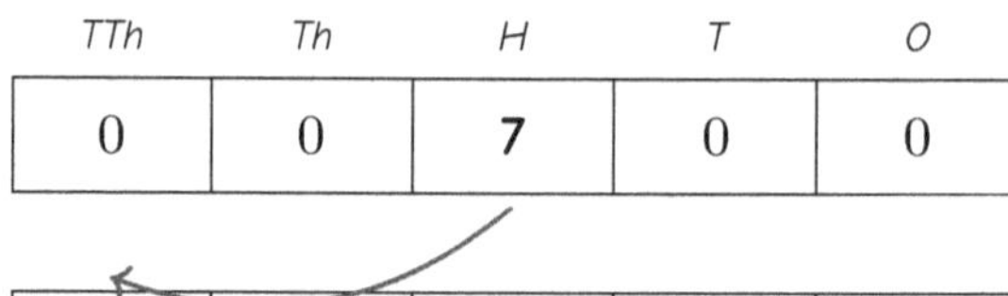

A) 374

B) 347

C) 743

D) 437

6. A stadium holds 72,500 fans. A second stadium holds 72,050 fans. Which stadium holds fewer fans?

A) First stadium

B) Second stadium

C) They hold equal fans

D) Cannot determine

7. Look at the place value chart below. The digit 7 moves from one position to another.

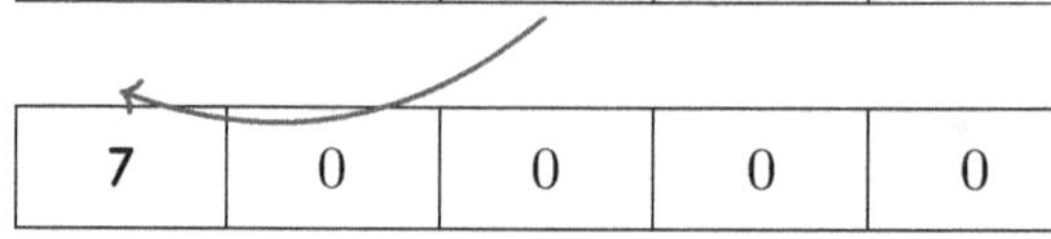

	TTh	Th	H	T	O
Before:	0	0	7	0	0
After:	7	0	0	0	0

How did the value of the digit 7 change?

A) It became 10 times bigger

B) It became 100 times bigger

C) It became 10 times smaller

D) It stayed the same

8. Estimate $671 - 289$ by rounding each number to the nearest hundred.

Your Answer:

9. *What is 7×905?*

  (A) 6,235   (B) 6,325

  (C) 6,335   (D) 6,435

10. *A rectangle is 71 feet wide and 84 feet long. What is the area of the rectangle?*

  (A) 5,854 *sq ft*   (B) 5,864 *sq ft*

  (C) 5,954 *sq ft*   (D) 5,964 *sq ft*

11. *A student says $\frac{5}{6} > \frac{7}{8}$ because $5 < 7$. Is the student correct?*

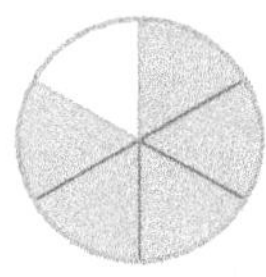

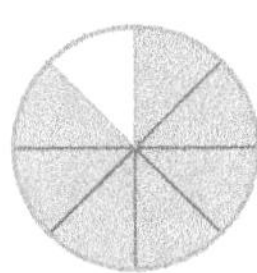

  (A) *Yes, because 5 is less than 7.*   (B) *No. $\frac{5}{6} < \frac{7}{8}$ because $\frac{5}{6} \approx 0.83$ and $\frac{7}{8} \approx 0.875$.*

  (C) *Yes, because $6 < 8$.*   (D) *No. $\frac{5}{6} = \frac{7}{8}$.*

12. $\frac{6}{10}$ can be broken into $\frac{4}{10} +$ ____.

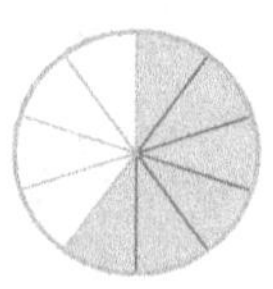

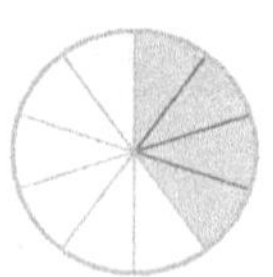

(A) $\frac{3}{10}$

(B) $\frac{2}{10}$

(C) $\frac{4}{10}$

(D) $\frac{6}{10}$

13. What is $\frac{4}{6} - \frac{1}{6}$?

(A) $\frac{3}{12}$

(B) $\frac{4}{6}$

(C) $\frac{3}{0}$

(D) $\frac{3}{6}$

14. What is $1\frac{2}{6} + 2\frac{3}{6}$?

(A) $3\frac{1}{6}$

(B) $3\frac{5}{12}$

(C) $4\frac{2}{6}$

(D) $3\frac{5}{6}$

15. Look at the number line below. It shows 4 equal jumps of $\frac{3}{6}$ starting from 0. Where does the last arrow land?

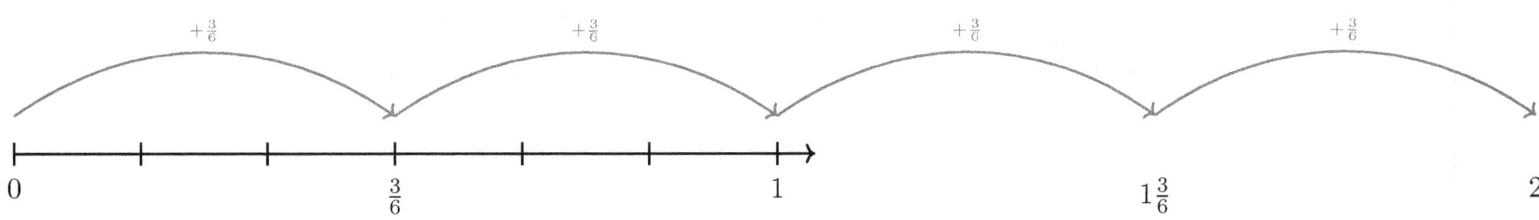

(A) $1\frac{2}{6}$

(B) $1\frac{3}{6}$

(C) 2

(D) $2\frac{3}{6}$

16. What is $\frac{2}{10} + \frac{35}{100}$?

Your Answer

17. A snack costs \$0.48. Write this price as a fraction of a dollar in two ways: once with a denominator of 100, and once simplified.

Your Answer

18. Compare: 0.6 ___ 0.43

(A) $<$

(B) $=$

(C) $>$

(D) Cannot be determined.

19. How many grams are in 1 kilogram?

(A) 100 g

(B) 10,000 g

(C) 10 g

(D) 1,000 g

Find more at
ViewMath.com/CT-Grade4

20. A dog weighs 5 kg. A cat weighs 3,000 g. Which animal is heavier?

(A) The cat

(B) The dog

(C) They weigh the same

(D) Cannot be determined

21. Look at the two rectangles below. What is their combined area?

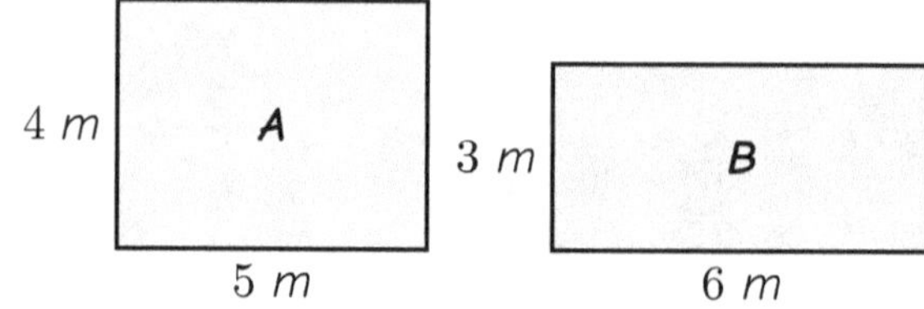

Your Answer:

22. A square has a perimeter of 48 inches. What is the length of one side?

(A) 4 in

(B) 8 in

(C) 12 in

(D) 24 in

23. A line plot shows: 2 X's at $\frac{1}{4}$, 3 X's at $\frac{1}{2}$, and 1 X at $\frac{3}{4}$. How many total data points are there?

(A) 3

(B) 6

(C) 8

(D) 12

Find more at
ViewMath.com/CT-Grade4

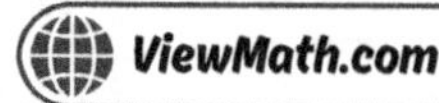

24. An angle that measures 55° is called a(n) _____________ angle.

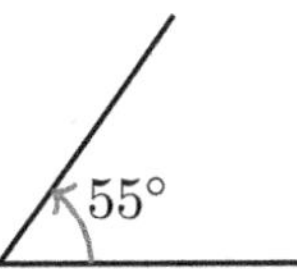

Your Answer:

25. Two adjacent angles share a side: one is 65° and the other is 25°. What is the total angle?

Your Answer:

26. How many endpoints does a **line** have?

(A) 0

(B) 1

(C) 2

(D) 4

27. Lena says that an angle of 91° is a right angle because it is "close to 90°." Is she correct?

(A) Yes, because it is almost 90°

(B) No, it is an acute angle

(C) No, it is an obtuse angle

(D) No, it is a straight angle

28. Which pair of lines below is **perpendicular**?

(A) Pair A   (B) Pair B

(C) Pair C   (D) Both Pair A and Pair C

29. A quadrilateral with **exactly one pair** of parallel sides is called a:

(A) Rectangle   (B) Parallelogram

(C) Trapezoid   (D) Rhombus

30. How many lines of symmetry does a square have?

Your Answer:

Find more at
ViewMath.com/CT-Grade4

 End of Practice Test 3 

Great job finishing the test!

☑ My Score

I got _____________ out of 30 questions right.

Check your answers in the **Answer Key** at the back of the book.

💡 Review any questions you missed. That's how we learn!

📊 Check Your Score Online!

Visit **ViewMath Academy** to enter your answers and see which topics you need to review. You can also explore lessons, take quizzes, track your scores, and save your progress!

viewmath.com/score/4.1.CT.03

Or go to viewmath.com/score and enter code: 4.1.CT.03

Answer Key & Explanations

> ⭐ **Check Your Answers!** ⭐
>
> First try each test on your own, then look here to check.
> Read the explanations to learn from any mistakes ⭐

✅ Practice Test 1 — Answer Key

| 1 | 35 | 2 | 15 | 3 | 42 or 49 | 4 | Prime; its only factors are 1 and 41. | 5 | C | 6 | C |

| 7 | B | 8 | B | 9 | 2,996 | 10 | B | 11 | C | 12 | B | 13 | C | 14 | B | 15 | C | 16 | B |

| 17 | B | 18 | C | 19 | B | 20 | 1,200 m (or 1 km 200 m) | 21 | C | 22 | B | 23 | C | 24 | C |

| 25 | B | 26 | B | 27 | D | 28 | False | 29 | A | 30 | C |

> 💡 **Time to Learn!** 💡
>
> Go through the explanations below, **especially for the questions you missed**.
>
> Understanding why each answer is correct makes you a stronger math thinker!
>
> 👍 **Tip:** Circle any questions you got wrong, then read their explanation carefully.

📖 Practice Test 1 — Detailed Explanations

1. $7 \times 5 = 35$ *pencils.*

2 Step 1: $45 \div 5 = 9$ students per team (not needed). There are 5 teams each needing 3 jump ropes: $5 \times 3 = 15$.

3 $7 \times 6 = 42$ and $7 \times 7 = 49$. Both are between 40 and 50.

4 Testing divisibility: 41 is not divisible by $2, 3, 5,$ or 7 (the primes up to $\sqrt{41} \approx 6.4$), so 41 is prime.

5 The hundred-thousands place is the leftmost digit of a six-digit number. In 307,416, the digit 3 is in the hundred-thousands place.

6 In 410,300 vs 410,030: compare hundreds: $3 > 0$. So $410,300 > 410,030$. The $<$ sign is wrong—this comparison is INCORRECT.

7 $3 \times 100 = 300$. Multiplying by 100 shifts all digits two places to the left (same as multiplying by 10 twice).

8 $387 \approx 400$ (tens digit $8 \geq 5$, round up) and $425 \approx 400$ (tens digit $2 < 5$, round down). $400 + 400 = 800$.

9 $7 \times 400 = 2{,}800$; $7 \times 20 = 140$; $7 \times 8 = 56$. Total: $2{,}800 + 140 + 56 = 2{,}996$.

10 The standard algorithm multiplies by the ones digit first (78×3), then by the tens digit (78×4, shifted one place left with a placeholder zero), and adds the two partial products.

11 All have the same numerator (1). The smaller the denominator, the larger the fraction. $\frac{1}{2}$ has the smallest denominator, so it is greatest.

12 $\frac{1}{8} + \frac{2}{8} + \frac{1}{8} = \frac{1+2+1}{8} = \frac{4}{8}$.

13 $5 - 2 = 3$; keep denominator 12: $\frac{3}{12}$.

14 Regroup: $6\frac{3}{5} = 5\frac{8}{5}$. Then $5 - 4 = 1$ and $\frac{8}{5} - \frac{4}{5} = \frac{4}{5}$. Answer: $1\frac{4}{5}$.

15 $6 \times 1 = 6$; $\frac{6}{4} = 1\frac{2}{4}$ because $6 \div 4 = 1$ R 2.

16 3 dimes $= \frac{3}{10} = \frac{30}{100}$. Add 8 pennies $= \frac{8}{100}$. Total: $\frac{30}{100} + \frac{8}{100} = \frac{38}{100}$.

17 0.25 is 25 hundredths $= \frac{25}{100}$.

18 $32.4 = 32.40$ (adding a zero at the end doesn't change the value). They tied.

19 The prefix kilo- means $1,000$, so 1 km $= 1,000$ m.

20 3 laps $\times 400$ m $= 1,200$ m. Since $1,000$ m $= 1$ km, the runner has covered 1 km and 200 m.

21 $A = 10 \times 10 = 100$ sq ft.

22 $P = 2(9) + 2(4) = 18 + 8 = 26$ cm.

23 When measuring in halves, the only value between 0 and 1 is $\frac{1}{2}$.

24 An obtuse angle is between $90°$ and $180°$. Only $110°$ fits that range.

25 $45 + 45 = 90°$, which is a **right** angle.

26 All three points are on the same line, so $\overline{CD}$ and $\overline{DE}$ are both parts (segments) of that one line.

27 In order from smallest to largest: acute $(< 90°)$, right $(= 90°)$, obtuse $(< 180°)$, straight $(= 180°)$. Straight is the largest.

28 Parallel lines never intersect, no matter how far they are extended.

Find more at
ViewMath.com/CT-Grade4

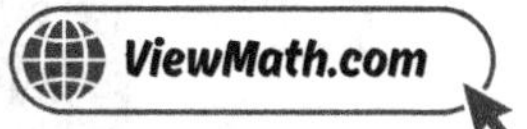

29 | A square is a special rectangle (4 right angles) that also has all 4 sides equal. It is also a special rhombus.

30 | An equilateral triangle has 3 lines of symmetry — one from each vertex to the midpoint of the opposite side.

✅ Practice Test 2 — Answer Key

1. $27; 3 \times 9 = 27$	2. B	3. 6 (or 8, 10, 14, 15)	4. C	5. D	6. C	7. 100 times	
8. C	9. 1,505	10. B	11. False	12. B	13. B	14. C	15. B	16. A
17. B	18. B	19. C	20. 5,500 g	21. A	22. C	23. C	24. B	25. C	26. D
27. Acute	28. B	29. B	30. 3				

💡 Time to Learn! 💡

Go through the explanations below, **especially for the questions you missed**.

Understanding why each answer is correct makes you a stronger math thinker!

👍 *Tip:* Circle any questions you got wrong, then read their explanation carefully.

📖 Practice Test 2 — Detailed Explanations

1. The array has 3 rows and 9 columns. $3 \times 9 = 27$ dots total.

2. Step 1: $7 \times 24 = 168$ crayons. Step 2: $168 - 15 = 153$ crayons.

Find more at
ViewMath.com/CT-Grade4

3) For example, factors of 6 are $1, 2, 3, 6$ — exactly 4 factors. Other valid answers include 8 $(1, 2, 4, 8)$, 10 $(1, 2, 5, 10)$, 14 $(1, 2, 7, 14)$, 15 $(1, 3, 5, 15)$.

4) $30 \div 3 = 10$ and $30 \div 5 = 6$. So 30 is a multiple of both 3 and 5.

5) In 850,293: 8 = hundred-thousands, 5 = ten-thousands. The value is $5 \times 10,000 = 50,000$.

6) Both towns have 183 in the thousands period. Compare hundreds: $4 > 0$. So $183,400 > 183,040$. Town A has more people.

7) Position A is the ten-thousands place $(9 \times 10,000 = 90,000)$. Position B is the hundreds place $(9 \times 100 = 900)$. $90,000 \div 900 = 100$. Position A is 100 times bigger.

8) The tens digit is 4. Look right at the ones digit: $7 \geq 5$, so round up. 4 becomes 5, giving 350.

9) 7×215: ones $7 \times 5 = 35$, write 5, carry 3; tens $7 \times 1 + 3 = 10$, write 0, carry 1; hundreds $7 \times 2 + 1 = 15$. Product is 1,505.

10) 36×52: $36 \times 2 = 72$ (ones row); $36 \times 50 = 1,800$ (tens row). Sum: $72 + 1,800 = 1,872$.

11) Convert $\frac{2}{3} = \frac{4}{6}$. Compare $\frac{4}{6}$ and $\frac{5}{6}$: $4 < 5$, so $\frac{2}{3} < \frac{5}{6}$.

12) The numerators must add to 7. Only $3 + 4 = 7$ works. $(4 + 4 = 8, 5 + 3 = 8, 2 + 6 = 8$ — all wrong.)

13) $\frac{5}{8} - \frac{2}{8} = \frac{3}{8}$. Check: $\frac{1}{8} + \frac{2}{8} = \frac{3}{8}$ ✓.

14) Whole numbers: $3 - 1 = 2$. Fractions: $\frac{7}{8} - \frac{5}{8} = \frac{2}{8}$. Answer: $2\frac{2}{8}$.

15) $5 \times 1 = 5$; $\frac{5}{3} = 1\frac{2}{3}$ because $5 \div 3 = 1$ R 2.

Find more at
ViewMath.com/CT-Grade4

16 $\frac{3}{10} = \frac{30}{100}$. The difference: $\frac{45}{100} - \frac{30}{100} = \frac{15}{100}$.

17 35 squares out of 100 are shaded. $\frac{35}{100} = 0.35$.

18 Add a zero: $14.5 = 14.50$. Compare 14.50 and 14.38: the hundredths show $50 > 38$, so $14.5 > 14.38$. Mia (with 14.38) was faster.

19 The pattern shows each cm $= 10$ mm. So $4 \times 10 = 40$ mm.

20 Convert: 4 kg $= 4{,}000$ g. Then $4{,}000 + 1{,}500 = 5{,}500$ g.

21 Floor area $= 9 \times 8 = 72$ sq m. Bookshelf area $= 2 \times 3 = 6$ sq m. Remaining: $72 - 6 = 66$ sq m.

22 Add all outer sides: $10 + 3 + 4 + 3 + 6 + 6 = 32$ ft.

23 $\frac{7}{8}$ inch is the largest value on the line plot.

24 One-quarter of $360°$ is $360 \div 4 = 90°$. A quarter turn makes a right angle.

25 $90 + 90 + 80 = 260$. Fourth angle $= 360 - 260 = 100°$.

26 A ray has one endpoint (a beginning) and goes on forever in one direction.

27 The angle opens less than $90°$ (less than a right angle), so it is acute.

28 The top edge and side edge of a door meet at a $90°$ angle, making them perpendicular.

29 "Quad" means four. A quadrilateral always has exactly 4 sides and 4 angles.

Find more at
ViewMath.com/CT-Grade4

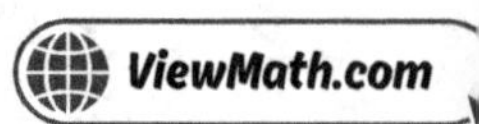

30 An equilateral triangle has 3 lines of symmetry, one from each vertex to the midpoint of the opposite side.

Practice Test 3 — Answer Key

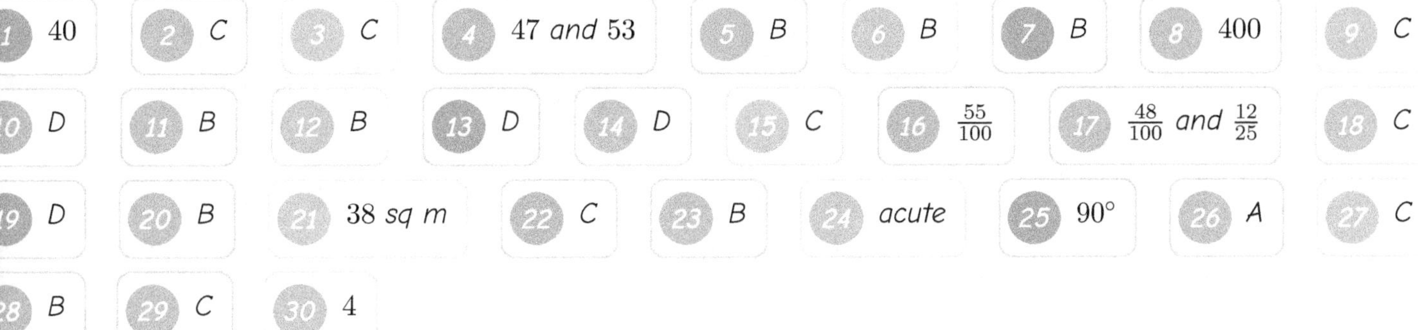

1 40	2 C	3 C	4 47 and 53	5 B	6 B	7 B	8 400	9 C
10 D	11 B	12 B	13 D	14 D	15 C	16 $\frac{55}{100}$	17 $\frac{48}{100}$ and $\frac{12}{25}$	18 C
19 D	20 B	21 38 sq m	22 C	23 B	24 acute	25 90°	26 A	27 C
28 B	29 C	30 4						

💡 Time to Learn! 💡

Go through the explanations below, **especially for the questions you missed**.

Understanding why each answer is correct makes you a stronger math thinker!

👍 **Tip:** Circle any questions you got wrong, then read their explanation carefully.

📖 Practice Test 3 — Detailed Explanations

1 8 times as many as 5 is $8 \times 5 = 40$.

2 Step 1: $48 \div 6 = 8$ bags baked. Step 2: $8 + 5 = 13$ bags total.

3 7 is prime — its only factor pair is $(1, 7)$. All others have more than one pair.

Find more at
ViewMath.com/CT-Grade4

(4) 47 has only factors 1 and 47. 53 has only factors 1 and 53. All other numbers between 45 and 55 are composite.

(5) There are 3 hundreds (300), 4 tens (40), and 7 ones (7). That makes $300 + 40 + 7 = 347$.

(6) Both start with 72. Compare hundreds: $5 > 0$. So $72,500 > 72,050$. The second stadium holds fewer fans.

(7) The 7 moved from the hundreds place (700) to the ten-thousands place (70,000). That's two places to the left: $700 \times 10 \times 10 = 70,000$, which is 100 times bigger.

(8) $671 \approx 700$ (tens digit $7 \geq 5$, round up) and $289 \approx 300$ (tens digit $8 \geq 5$, round up). $700 - 300 = 400$.

(9) Ones: $7 \times 5 = 35$, write 5, carry 3. Tens: $7 \times 0 + 3 = 3$. Hundreds: $7 \times 9 = 63$. Product is 6,335.

(10) 71×84: $70 \times 80 = 5,600$, $70 \times 4 = 280$, $1 \times 80 = 80$, $1 \times 4 = 4$. Sum: $5,600 + 280 + 80 + 4 = 5,964$ sq ft.

(11) You cannot compare numerators when denominators differ. Using a common denominator of 24: $\frac{5}{6} = \frac{20}{24}$ and $\frac{7}{8} = \frac{21}{24}$. So $\frac{5}{6} < \frac{7}{8}$.

(12) $4 + ? = 6$, so $? = 2$. The missing piece is $\frac{2}{10}$.

(13) Subtract the numerators, keep the denominator. $4 - 1 = 3$, so $\frac{4}{6} - \frac{1}{6} = \frac{3}{6}$.

(14) Whole numbers: $1 + 2 = 3$. Fractions: $\frac{2}{6} + \frac{3}{6} = \frac{5}{6}$. Answer: $3\frac{5}{6}$.

(15) $4 \times \frac{3}{6} = \frac{12}{6} = 2$. Each jump is $\frac{3}{6}$ $(= \frac{1}{2})$, and 4 jumps of $\frac{1}{2}$ lands on 2.

(16) $\frac{2}{10} = \frac{20}{100}$. Then $\frac{20}{100} + \frac{35}{100} = \frac{55}{100}$.

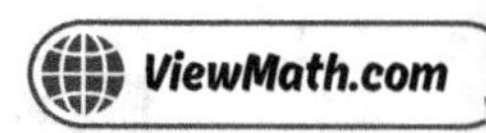

17. $0.48 = \frac{48}{100}$. Simplified by dividing by 4: $\frac{48 \div 4}{100 \div 4} = \frac{12}{25}$.

18. Add a zero: $0.6 = 0.60$. Compare 0.60 and 0.43: the tenths digit $6 > 4$, so $0.6 > 0.43$.

19. The prefix kilo- means $1,000$, so $1\ kg = 1,000\ g$.

20. Convert: $5\ kg = 5,000\ g$. Since $5,000 > 3,000$, the dog is heavier.

21. Rectangle A: $5 \times 4 = 20$ sq m. Rectangle B: $6 \times 3 = 18$ sq m. Total: $20 + 18 = 38$ sq m.

22. A square has 4 equal sides: $48 \div 4 = 12$ in per side.

23. $2 + 3 + 1 = 6$ total data points.

24. $55°$ is less than $90°$, so it is an acute angle.

25. $65 + 25 = 90°$.

26. A line goes on forever in both directions and has no endpoints at all.

27. A right angle is EXACTLY $90°$. Since $91°$ is slightly more than $90°$, it is obtuse.

28. Perpendicular lines meet at a $90°$ angle (shown by the small square). Only Pair B has a right angle mark.

29. A trapezoid has exactly 1 pair of parallel sides. A parallelogram, rectangle, rhombus, and square all have 2 pairs.

30. A square has 4 lines of symmetry: one vertical, one horizontal, and two diagonal.

Find more at
ViewMath.com/CT-Grade4

Great job checking your work!

Keep practicing and you'll be a math star!

Find more at
ViewMath.com/CT-Grade4

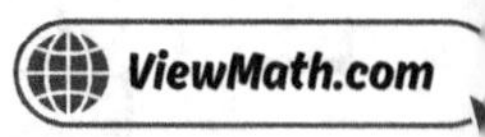